3 Virginia SOL Grade 3 Math Practice Tests

Full-Length Test Prep with Detailed Answer Explanations

Dr. A. Nazari

Copyright © 2026 Dr. A. Nazari

Published by View Math Education

ViewMath.com

3 Practice Tests to Get You Started!

Hey there, future math whiz! ⭐

This book has **3 full practice tests** to help you warm up for the real thing. Think of it like stretching before a big game — these tests will get your brain ready and show you what to expect!

👍 Three tests is the **perfect start**!

👍 Each one helps you feel **more ready**!

👍 You'll be surprised how much you **already know**!

Sharpen your pencil and let's get warmed up! 

> 66 Three practice tests is a great way to start. Take your time with each one, and you'll feel more confident every step of the way! 99

📖 How to Use This Book 📖

Your quick-start guide to 3 great practice tests!

☰ What's Inside This Book

- **3 Full-Length Practice Tests** — Each one covers all the Grade 3 math topics you need to know!
- **Answer Key with Explanations** — Find out why each answer is correct, not just what the answer is.
- **Reference Pages** — A math symbols chart and multiplication table you can peek at any time.
- **A Test Tracker** — Write down your scores and watch your confidence grow!

📅 A Simple 3-Test Plan

With just 3 tests, here's a great way to use them:

- **Test 1** — **The Warm-Up.** Take this test without a timer. Get comfortable with the question types. Don't worry about your score — just do your best!
- **Test 2** — **The Practice Round.** After reviewing Test 1, try this one with a timer (ask a grown-up!). Focus on the topics that were tricky last time.
- **Test 3** — **The Real Deal.** Treat this like the actual test: quiet room, timed, no peeking at answers. See how much you've improved!

⬤ Multiple Choice

Pick the **one best answer** from choices A, B, C, or D. Not sure? Cross out the ones you know are wrong, then pick from what's left. That's a smart move!

✏ Short Answer

Write your answer **and** show your work! Even if your final answer isn't right, showing your steps can earn you credit. Use scratch paper if you need more room.

66 After Each Test 99

Flip to the Answer Key and check your work. For every question you got wrong, **read the explanation carefully.** Then write the tricky topics on your Test Tracker page. If you need extra help, grab our **Grade 3 Math Study Guide!**

Find more at
ViewMath.com/VA-Grade3

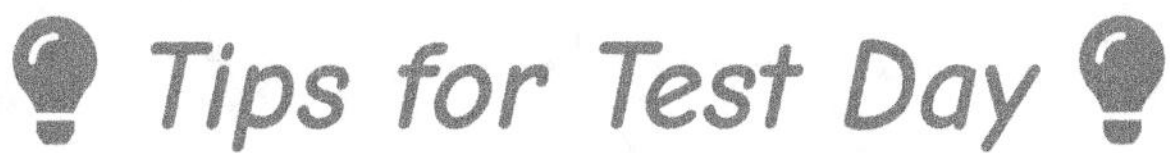

💡 Tips for Test Day 💡

Easy tricks to help you feel calm and do your best!

🌙 The Night Before

- ✅ **Sleep early** — *your brain learns while you sleep!*
- ✅ **Pack your supplies** — *pencils, eraser, scratch paper, all ready to go.*
- ✅ **Tell yourself:** *"I've been practicing. I'm going to do great!"*

👍 5 Simple Rules for Every Test

1. **Read the question twice.** *The first time to understand it. The second time to catch details.*
2. **Show your work.** *Write the steps down, even on scratch paper. It helps you think!*
3. **Skip the hard ones.** *Put a small star next to tricky questions and come back later. Answer the easy ones first!*
4. **Never leave a blank.** *For multiple choice, your best guess is better than no answer at all.*
5. **Check your work.** *Finished early? Go back and re-read your answers.*

✅ Smart Moves

- Take a deep breath before you begin
- Underline key words in the question
- Use drawings or number lines to help
- Cross out wrong answers first
- Double-check addition and subtraction

❌ Traps to Avoid

- Rushing and not reading carefully
- Picking the first answer that "looks right"
- Forgetting to carry or borrow numbers
- Skipping a question permanently
- Panicking when you see a tough problem

❝ Remember, the very first practice test is the hardest — not because the questions are harder, but because everything is new! By Test 3, you'll feel like a pro. Trust me! ❞

Get Ready to Practice

Here's everything you need before you start!

Pencils

Sharpened and ready!

Eraser

Everyone makes mistakes!

Scratch Paper

For working things out

A Calm Spot

Somewhere quiet to focus

A Grown-Up

To help set a timer

A Can-Do Attitude

You've totally got this!

✓ Allowed During Tests

- Pencils and erasers
- Blank scratch paper
- The **reference pages** in this book
- A ruler (for measurement questions)

✗ Not Allowed

- Calculators
- Phones, tablets, or computers
- Help from anyone else
- Your study guide (save it for after!)

👥 For Parents & Teachers

- With only 3 tests, **space them at least a week apart**. This gives time to review mistakes before trying the next one.
- Let your child take Test 1 untimed to build familiarity.
- After each test, go through the Answer Key together. Focus on **understanding the "why,"** not just the score.
- If a topic keeps tripping them up, review it in our **Grade 3 Math Study Guide** before the next practice test.
- Celebrate every bit of progress — even getting one more question right is a win!

$\mathbf{X}^1$ Math Reference Sheet $\mathbf{X}^1$

You may use this page during your practice tests!

Symbol	Name	What It Means	
$+$	Plus (Add)	Put numbers together.	$3 + 5 = 8$
$-$	Minus (Subtract)	Take away from a number.	$9 - 4 = 5$
$\times$	Times (Multiply)	Add equal groups.	$4 \times 3 = 12$
$\div$	Divide	Split into equal groups.	$12 \div 3 = 4$
$=$	Equals	Both sides are the same.	$2 + 3 = 5$
$>$	Greater Than	The left number is bigger.	$7 > 3$
$<$	Less Than	The left number is smaller.	$2 < 9$
$\frac{1}{2}$	Fraction Bar	Part of a whole.	$\frac{1}{2}$ means 1 out of 2 equal parts

📘 Key Math Words

- **Sum** — the answer when you add
- **Difference** — the answer when you subtract
- **Product** — the answer when you multiply
- **Quotient** — the answer when you divide
- **Factor** — a number you multiply
- **Array** — objects in rows and columns
- **Fraction** — a part of a whole
- **Numerator** — the top number in a fraction
- **Denominator** — the bottom number
- **Equation** — a math sentence with $=$
- **Estimate** — a smart guess, close to the real answer
- **Perimeter** — the distance around a shape
- **Area** — the space inside a shape
- **Rounding** — making a number simpler by going to the nearest ten or hundred

🔍 Word Problem Clue Words

- **Add** (+): in all, total, altogether, combined, sum, both, more
- **Subtract** (−): how many more, how many left, fewer, difference, remain
- **Multiply** (×): each, every, groups of, times, rows of, per
- **Divide** (÷): share equally, split, each group, how many groups, per

Find more at
ViewMath.com/VA-Grade3

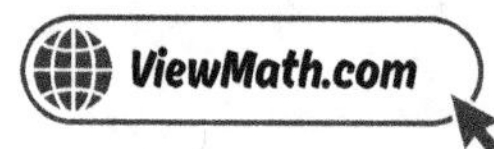

Multiplication Table

You may use this table during your practice tests!

×	1	2	3	4	5	6	7	8	9	10	11
1	1	2	3	4	5	6	7	8	9	10	11
2	2	4	6	8	10	12	14	16	18	20	22
3	3	6	9	12	15	18	21	24	27	30	33
4	4	8	12	16	20	24	28	32	36	40	44
5	5	10	15	20	25	30	35	40	45	50	55
6	6	12	18	24	30	36	42	48	54	60	66
7	7	14	21	28	35	42	49	56	63	70	77
8	8	16	24	32	40	48	56	64	72	80	88
9	9	18	27	36	45	54	63	72	81	90	99
10	10	20	30	40	50	60	70	80	90	100	110
11	11	22	33	44	55	66	77	88	99	110	121

How to Use This Table

To find **4 × 7**:

1. Find **4** in the left column (blue).
2. Find **7** in the top row (blue).
3. Follow the row and column until they meet: the answer is **28**!

📈 My Confidence Tracker 📈

Record your scores below. You'll be amazed at your progress!

My name: _________________________________

☑ Test	📅 Date	★ Score	☺ How I Feel
1		/	☺
2		/	☺
3		/	☺

The easiest topic for me was:

____

The trickiest topic for me was:

____

One thing I got better at from Test 1 to Test 3:

____

Next time I want to try:

____

 *You just finished 3 practice tests — that's awesome! Compare your first score to your last. I bet you'll see real improvement. Ready for more? Check out our 5-test or 7-test books for even more practice!*

Get Online

Find more at
ViewMath.com/VA-Grade3

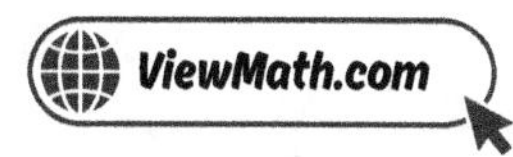

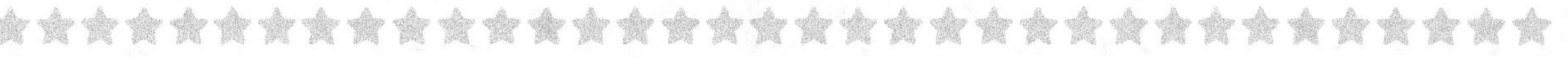

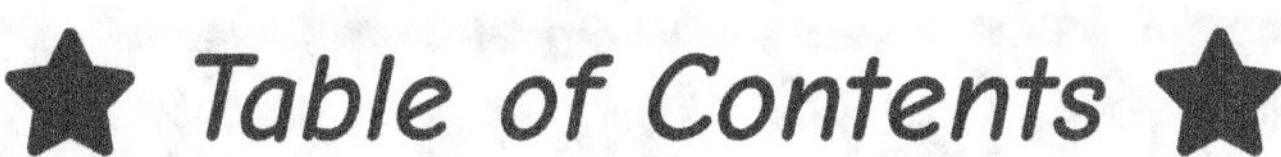 Table of Contents

Here's what we'll explore together!

 Let's learn and have fun!

Practice Test 1

30 Questions

✏ Before You Start ✏

- ✓ **Read each question carefully** before choosing your answer.
- ✓ **Show your work** on scratch paper when you need to.
- ✓ **Skip hard questions** and come back to them later.
- ✓ **Check your answers** when you're done.
- ✓ **Take your time** — there's no rush!

★ You've Got This! ★

Do your best and show what you know!

1. *Write 5,600 in expanded form.*

   Your Answer:

2. *What is 2,500 rounded to the nearest 1,000?*

   (A) 2,000

   (B) 2,500

   (C) 3,000

   (D) 3,500

3. *Which group contains only even numbers?*

   (A) 12, 34, 57

   (B) 20, 46, 88

   (C) 31, 50, 72

   (D) 14, 63, 90

4. *A school raised $8,000 last year and $2,000 this year. How much did the school raise in total?*

   (A) $6,000

   (B) $10,000

   (C) $82,000

   (D) $28,000

5. *What is 8,912 rounded to the nearest 1,000?*

   (A) 8,000

   (B) 8,900

   (C) 9,000

   (D) 10,000

6. *What is $198 + 345 + 267$?*

   Your Answer:

Find more at
ViewMath.com/VA-Grade3

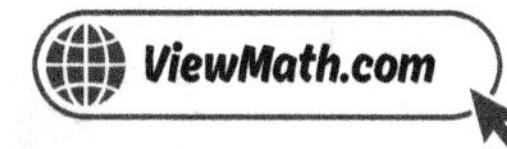

7. *What is $6{,}789 + 3{,}456$?*

Your Answer:

8. *When you estimate $347 + 582$ by rounding each number to the nearest hundred, the estimate is 900. The exact sum is 929. Which statement is true?*

(A) *The estimate is greater than the exact sum*

(B) *The estimate is less than the exact sum*

(C) *The estimate equals the exact sum*

(D) *You cannot compare an estimate to an exact answer*

9. *I am a number. When you divide me by 8, you get 9. What number am I?*

Your Answer:

10. *Find the missing number. $72 \div ? = 8$*

(A) 7

(B) 8

(C) 9

(D) 10

11. *Find a fraction equivalent to $\frac{1}{3}$ by multiplying the numerator and denominator by 2.*

(A) $\frac{2}{3}$

(B) $\frac{1}{6}$

(C) $\frac{2}{6}$

(D) $\frac{3}{6}$

12. *Which fraction does NOT equal 1?*

(A) $\frac{2}{2}$

(B) $\frac{8}{8}$

(C) $\frac{3}{6}$

(D) $\frac{4}{4}$

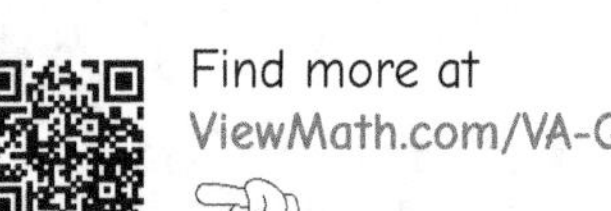

Find more at
ViewMath.com/VA-Grade3

13. Is $\frac{2}{6}$ more or less than $\frac{1}{2}$?

> *Your Answer:*

14. A recipe needs $\frac{3}{6}$ cup of sugar and $\frac{2}{6}$ cup of brown sugar. How much sugar is needed in total?

   (A) $\frac{1}{6}$ cup   (B) $\frac{5}{12}$ cup

   (C) $\frac{5}{6}$ cup   (D) $\frac{6}{6}$ cup

15. What is $\frac{4}{6} - \frac{2}{6}$?

> *Your Answer:*

16. What is $3\frac{1}{3}$ written as an improper fraction?

   (A) $\frac{4}{3}$ (B) $\frac{7}{3}$

   (C) $\frac{10}{3}$ (D) $\frac{31}{3}$

17. A clock shows the short hand exactly on the 12 and the long hand on the 12. What time is it?

   (A) 12:12   (B) 12:00

   (C) 0:00   (D) 6:00

18. How many half inches are in 3 inches?

> *Your Answer:*

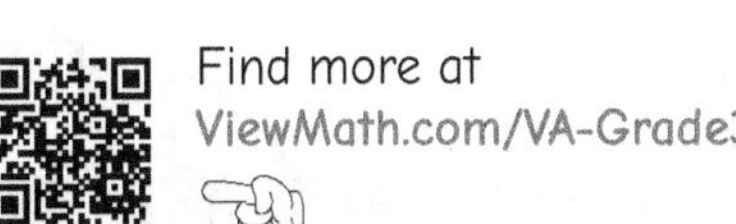

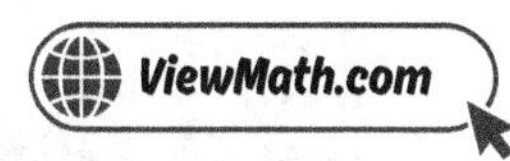

19. A box of cereal has a mass of 500 g. You buy 2 boxes. What is the total mass?

   (A) 502 g

   (B) 700 g

   (C) 1,000 g

   (D) 5,000 g

20. Lily has $6.00. She buys a sticker pack for $2.35 and a pen for $1.80. Does she have enough money left to buy a snack for $2.00?

   (A) Yes, she has $2.85 left

   (B) Yes, she has $2.00 left

   (C) No, she only has $1.85 left

   (D) No, she only has $1.65 left

21. Which is warmer: 45°F or 72°F?

   (A) 45°F

   (B) 72°F

   (C) They are the same.

   (D) Cannot tell.

22. A picture graph shows that the row for Red has 8 stars and the row for Green has 5 stars. Each star stands for 2 crayons. How many crayons are there in all?

   Your Answer

23. What does a line plot use to show data?

   (A) Colored bars

   (B) Little pictures in rows

   (C) X marks above a number line

   (D) Lines connecting dots

Find more at
ViewMath.com/VA-Grade3

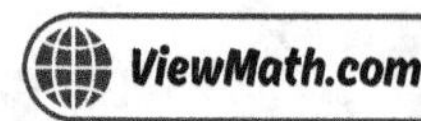
ViewMath.com

24. You flip a coin. What is the chance of getting heads?

(A) Certain

(B) Equally likely

(C) Unlikely

(D) Impossible

25. What is the top point of a pyramid called?

(A) Base

(B) Edge

(C) Vertex

(D) Apex

26. What is a unit square?

(A) A square with sides that are 10 units long

(B) A square with sides that are 1 unit long

(C) Any square shape

(D) A rectangle with 4 equal sides

27. A rectangle has an area of 42 sq cm and a length of 7 cm. What is the width?

Your Answer:

28. What type of angle is the corner of a book?

(A) Acute angle

(B) Right angle

(C) Obtuse angle

(D) Straight angle

29. A nonagon has how many vertices?

(A) 7

(B) 8

(C) 9

(D) 10

Find more at
ViewMath.com/VA-Grade3

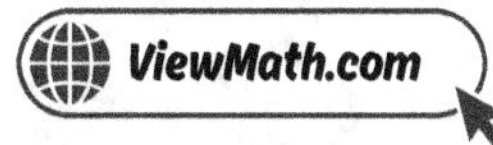

30. Two circles have different radii. One has radius 2 cm and the other has radius 4 cm. Are they congruent?

(A) Yes, because they are both circles.

(B) Yes, because circles are always congruent.

(C) No, because they are different sizes.

(D) No, because circles cannot be congruent.

 # End of Practice Test 1 

Great job finishing the test!

 My Score

I got __________ out of 30 questions right.

*Check your answers in the **Answer Key** at the back of the book.*

Review any questions you missed. That's how we learn!

📊 Check Your Score Online!

Visit **ViewMath Academy** to enter your answers and see which topics you need to review. You can also explore lessons, take quizzes, track your scores, and save your progress!

viewmath.com/score/3.1.VA.01

Or go to viewmath.com/score and enter code: 3.1.VA.01

2

Practice Test 2

☑ *30 Questions*

✏ Before You Start ✏

- ✓ **Read each question carefully** before choosing your answer.
- ✓ **Show your work** on scratch paper when you need to.
- ✓ **Skip hard questions** and come back to them later.
- ✓ **Check your answers** when you're done.
- ✓ **Take your time** — there's no rush!

⭐ You've Got This! ⭐

Do your best and show what you know!

1. Write the expanded form of 9,205.

   Your Answer:

2. What is 6,782 rounded to the nearest 1,000?

   (A) 6,000   (B) 6,700

   (C) 6,800   (D) 7,000

3. List all the even numbers between 21 and 30.

   Your Answer:

4. How many hundreds are in 10,000?

   Your Answer:

5. What is 7,777 rounded to the nearest 1,000?

   (A) 7,000   (B) 7,700

   (C) 7,800   (D) 8,000

6. A library has 583 fiction books and 268 nonfiction books. How many books does the library have in all?

   Your Answer:

Find more at
ViewMath.com/VA-Grade3

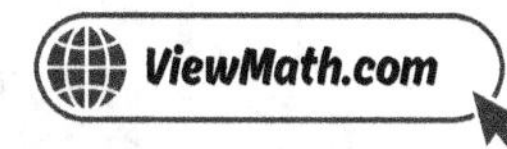

7. When adding $2{,}847 + 5{,}396$, which column do you start with?

   (A) Thousands   (B) Hundreds

   (C) Tens   (D) Ones

8. Estimate $3{,}412 + 2{,}678$ by rounding each number to the nearest thousand.

   (A) 5,000   (B) 6,000

   (C) 7,000   (D) 6,090

9. What is $72 \div 9$?

   (A) 7   (B) 8

   (C) 9   (D) 63

10. Find the missing number. $? \times 7 = 63$

    (A) 7   (B) 8

    (C) 9   (D) 10

11. $\frac{1}{2} = \frac{?}{6}$. What is the missing number?

    (A) 1   (B) 2

    (C) 3   (D) 4

12. What whole number does $\frac{9}{3}$ equal?

    (A) 1   (B) 3

    (C) 6   (D) 9

Find more at
ViewMath.com/VA-Grade3

ViewMath.com

13. Which fraction is closest to 1?

(A) $\frac{1}{4}$

(B) $\frac{2}{4}$

(C) $\frac{3}{4}$

(D) $\frac{1}{8}$

14. Alex walked $\frac{3}{8}$ of a mile in the morning and $\frac{4}{8}$ of a mile in the afternoon. How far did he walk total?

(A) $\frac{1}{8}$ mile

(B) $\frac{7}{16}$ mile

(C) $\frac{7}{8}$ mile

(D) $\frac{12}{8}$ mile

15. When you subtract fractions with the same denominator, what do you do?

(A) Subtract both the numerators and the denominators.

(B) Subtract the numerators and keep the denominator the same.

(C) Subtract the denominators and keep the numerator the same.

(D) Multiply the numerators.

16. A recipe needs $2\frac{1}{2}$ cups of milk. What is this as an improper fraction?

(A) $\frac{3}{2}$

(B) $\frac{4}{2}$

(C) $\frac{5}{2}$

(D) $\frac{21}{2}$

17. Which time does "half past 4" mean?

(A) 4:00

(B) 4:15

(C) 4:30

(D) 4:45

18. A marker is $4\frac{1}{2}$ inches long. A pen is 6 inches long. How much longer is the pen than the marker?

Your Answer:

19. Convert 3 kilograms to grams.

Your Answer:

20. *Add:* $6.40 + $2.35

Your Answer:

21. Which temperature would be best for a hot summer day at the beach?

(A) 30°F

(B) 50°F

(C) 65°F

(D) 92°F

22. Lily says the picture graph below shows that 4 kids were on the swings.

Swings: ★★★★

Slide: ★★★★★★

Key: Each ★ = 2 kids

What is the correct number of kids on the swings?

(A) 2

(B) 4

(C) 6

(D) 8

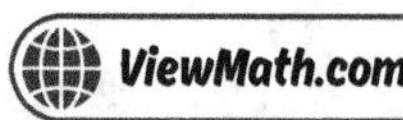

23. Which type of graph is BEST for showing measurement data like the lengths of worms in inches?

(A) Picture graph

(B) Bar graph

(C) Line plot

(D) Tally chart

24. A spinner has 8 equal sections. 6 are green and 2 are white. Is landing on green likely or unlikely?

Your Answer:

25. Which 3D shape has 0 faces, 0 edges, and 0 vertices?

(A) Cone

(B) Cylinder

(C) Cube

(D) Sphere

26. Alex counted 5 rows and 6 columns of unit squares in a rectangle. What is the area?

(A) 11 square units

(B) 22 square units

(C) 30 square units

(D) 56 square units

27. A rectangle is 6 cm long and 8 cm wide. What is its area?

Your Answer:

28. A right angle measures exactly how many degrees?

Your Answer:

Find more at
ViewMath.com/VA-Grade3

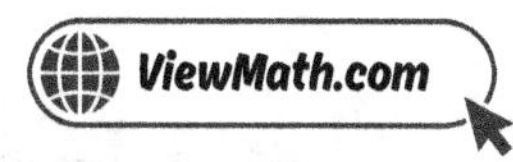

29. Which is the correct order of polygons from fewest sides to most sides?

(A) Pentagon, quadrilateral, hexagon, heptagon

(B) Triangle, quadrilateral, pentagon, hexagon

(C) Hexagon, octagon, nonagon, pentagon

(D) Quadrilateral, triangle, pentagon, hexagon

30. Can two shapes be congruent if one is flipped upside down? Explain.

Your Answer:

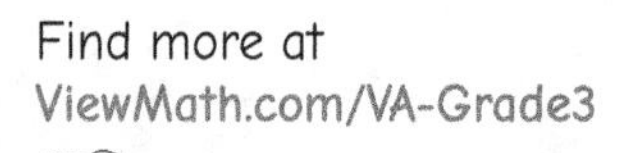

 End of Practice Test 2 

Great job finishing the test!

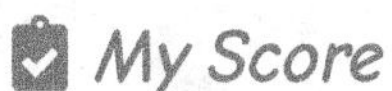 **My Score**

I got ___________ out of 30 questions right.

*Check your answers in the **Answer Key** at the back of the book.*

Review any questions you missed. That's how we learn!

📊 Check Your Score Online!

Visit **ViewMath Academy** to enter your answers and see which topics you need to review. You can also explore lessons, take quizzes, track your scores, and save your progress!

viewmath.com/score/3.1.VA.02

Or go to viewmath.com/score and enter code: 3.1.VA.02

Practice Test 3

📋 *30 Questions*

✏️ Before You Start ✏️

- ✔ **Read each question carefully** before choosing your answer.
- ✔ **Show your work** on scratch paper when you need to.
- ✔ **Skip hard questions** and come back to them later.
- ✔ **Check your answers** when you're done.
- ✔ **Take your time** — there's no rush!

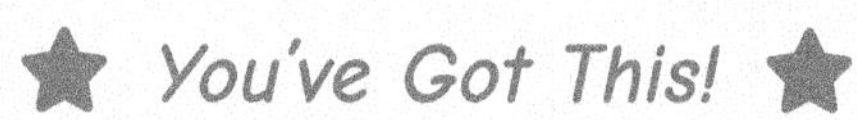

★ You've Got This! ★

Do your best and show what you know!

1. Each place value is how many times bigger than the place to its right?

(A) 2 times

(B) 5 times

(C) 10 times

(D) 100 times

2. A store sold 423 books this month. About how many books is that, rounded to the nearest 100?

Your Answer:

3. If you add two odd numbers, the result is always:

(A) Odd

(B) Even

(C) Greater than 10

(D) An odd number less than 20

4. What number is 1 more than 9,999?

(A) 9,000

(B) 9,990

(C) 10,000

(D) 10,001

5. Round 5,500 to the nearest 1,000.

Your Answer:

6. Which addition problem requires regrouping?

(A) $135 + 243$

(B) $428 + 351$

(C) $367 + 485$

(D) $512 + 136$

7. What is $2{,}103 + 5{,}642$?

(A) 7,745

(B) 7,645

(C) 7,755

(D) 8,745

8. Estimate $724 - 389$ by rounding each number to the nearest hundred.

(A) 200

(B) 300

(C) 400

(D) 335

9. What is $35 \div 5$?

(A) 5

(B) 6

(C) 7

(D) 8

10. Find the missing number. $8 \times ? = 48$

(A) 5

(B) 6

(C) 7

(D) 8

11. Which fraction is NOT equivalent to $\frac{2}{3}$?

(A) $\frac{4}{6}$

(B) $\frac{6}{8}$

(C) $\frac{6}{9}$

(D) $\frac{8}{12}$

12. What whole number does $\frac{10}{2}$ equal?

Your Answer:

13. Ruby says $\frac{1}{8} > \frac{1}{3}$ because $8 > 3$. Is Ruby correct?

(A) Yes, bigger numbers are always greater

(B) No, $\frac{1}{3} > \frac{1}{8}$ because thirds are bigger pieces than eighths

(C) They are equal

(D) Yes, eighths are bigger than thirds

14. What is $\frac{1}{6} + \frac{5}{6}$?

(A) $\frac{6}{12}$

(B) $\frac{5}{6}$

(C) $\frac{6}{6}$

(D) $\frac{4}{6}$

15. Ben has $\frac{6}{6}$ of a chocolate bar. He gives $\frac{4}{6}$ to his friend. How much does he have left?

Your Answer:

16. Which of the following is a proper fraction?

(A) $\frac{5}{3}$

(B) $\frac{6}{6}$

(C) $\frac{8}{4}$

(D) $\frac{2}{3}$

17. A clock shows the short hand past the 3 and the long hand on the 9. What time is it?

(A) 9:03

(B) 9:15

(C) 3:09

(D) 3:45

18. Max started measuring his pencil at the 1-inch mark. The other end reached the 7-inch mark. He says the pencil is 7 inches long. What is the real length?

(A) 7 inches

(B) 6 inches

(C) 8 inches

(D) 1 inch

19. 9,000 grams = how many kilograms?

(A) 9 kg

(B) 90 kg

(C) 900 kg

(D) 9,000 kg

20. Jake has $7.50. He buys a toy car for $4.25. How much money does he have left?

(A) $2.25

(B) $3.25

(C) $3.75

(D) $4.25

21. It was 80°F during the day. At night the temperature dropped 25 degrees. What is the nighttime temperature?

Your Answer:

22. What does the **key** on a picture graph tell you?

(A) The title of the graph

(B) How many categories there are

(C) What each symbol stands for

(D) Who made the graph

23. A line plot shows pencil lengths. There are 3 X marks above $1\frac{1}{2}$ inches. What does this mean?

(A) 3 pencils are $1\frac{1}{2}$ inches long

(B) $1\frac{1}{2}$ pencils are 3 inches long

(C) The total length is $4\frac{1}{2}$ inches

(D) There are 3 pencils in all

Find more at
ViewMath.com/VA-Grade3

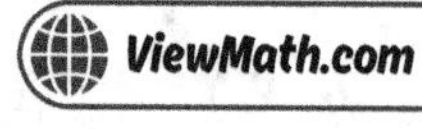

24. A jar has 6 green candies and 6 red candies. If you pick one without looking, what is the chance of getting green?

(A) Certain

(B) Equally likely

(C) Unlikely

(D) Impossible

25. Name a real-world object shaped like each: (a) a rectangular prism, (b) a sphere, (c) a cylinder.

Your Answer:

26. A garden has 7 rows and 3 columns of square tiles. How many square tiles cover the garden?

(A) 10 tiles

(B) 18 tiles

(C) 21 tiles

(D) 24 tiles

27. Which rectangle has the smallest area?

(A) 6 cm × 5 cm

(B) 3 cm × 8 cm

(C) 4 cm × 7 cm

(D) 2 cm × 10 cm

28. Carlos says a line segment goes on forever. Is he correct?

(A) Yes, line segments go on forever in both directions.

(B) Yes, line segments go on forever in one direction.

(C) No, a line segment has two endpoints and stops.

(D) No, a line segment has no endpoints.

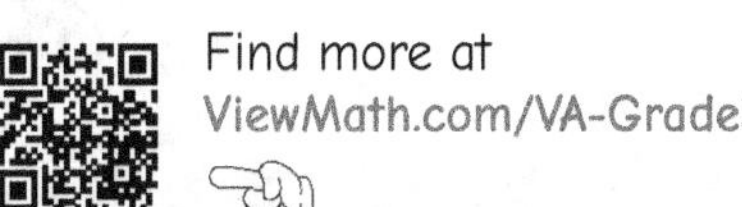
Find more at
ViewMath.com/VA-Grade3

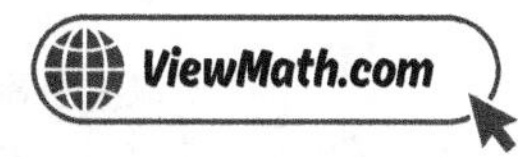

29. A polygon has 3 more sides than a pentagon. What polygon is it?

(A) Hexagon

(B) Heptagon

(C) Octagon

(D) Nonagon

30. A triangle has sides 3 cm, 4 cm, and 5 cm. Another has sides 3 cm, 4 cm, and 6 cm. Are they congruent?

(A) Yes, because two sides match.

(B) Yes, because they are both triangles.

(C) No, because all side lengths must match.

(D) No, because triangles are never congruent.

 End of Practice Test 3 

Great job finishing the test!

 My Score

I got ___________ out of 30 questions right.

Check your answers in the **Answer Key** at the back of the book.

 Review any questions you missed. That's how we learn!

Check Your Score Online!

Visit **ViewMath Academy** to enter your answers and see which topics you need to review. You can also explore lessons, take quizzes, track your scores, and save your progress!

viewmath.com/score/3.1.VA.03

Or go to viewmath.com/score and enter code: 3.1.VA.03

Answer Key & Explanations

⭐ Check Your Answers! ⭐

First try each test on your own, then look here to check.

Read the explanations to learn from any mistakes ⭐

✅ Practice Test 1 — Answer Key

1 $5{,}000 + 600$ | **2** C | **3** B | **4** B | **5** C | **6** 810 | **7** 10,245 | **8** B

9 72 | **10** C | **11** C | **12** C | **13** Less | **14** C | **15** $\dfrac{2}{6}$ | **16** C | **17** B | **18** 6

19 C | **20** C | **21** B | **22** 26 | **23** C | **24** B | **25** D | **26** B | **27** 6 cm | **28** B

29 C | **30** C

💡 Time to Learn! 💡

Go through the explanations below, **especially for the questions you missed**.

Understanding why each answer is correct makes you a stronger math thinker!

👍 **Tip:** Circle any questions you got wrong, then read their explanation carefully.

📖 Practice Test 1 — Detailed Explanations

1 $5{,}600 = 5{,}000 + 600 + 0 + 0.$ The tens and ones are both 0.

2 The hundreds digit is 5. Since $5 \geq 5$, round up. $2{,}500 \approx 3{,}000$.

3 20 (ends in 0), 46 (ends in 6), 88 (ends in 8) are all even. The other groups each contain at least one odd number.

4 $8{,}000 + 2{,}000 = 10{,}000$.

5 The hundreds digit is 9. Since $9 \geq 5$, we round up. $8{,}912 \approx 9{,}000$.

6 First add $198 + 345 = 543$. Ones: $8 + 5 = 13$, carry 1. Tens: $9 + 4 + 1 = 14$, carry 1. Hundreds: $1 + 3 + 1 = 5$. Then $543 + 267 = 810$. Ones: $3 + 7 = 10$, carry 1. Tens: $4 + 6 + 1 = 11$, carry 1. Hundreds: $5 + 2 + 1 = 8$.

7 Ones: $9 + 6 = 15$, carry 1. Tens: $8 + 5 + 1 = 14$, carry 1. Hundreds: $7 + 4 + 1 = 12$, carry 1. Thousands: $6 + 3 + 1 = 10$. The sum is 10,245 — a 5-digit number! Two 4-digit numbers can add up to more than 9,999.

8 $347 \approx 300$ and $582 \approx 600$. The estimate is $300 + 600 = 900$, which is less than the exact sum of 929.

9 If the number $\div 8 = 9$, then the number $= 9 \times 8 = 72$.

10 Think: $8 \times ? = 72$. Since $8 \times 9 = 72$, the missing number is 9.

11 $\frac{1 \times 2}{3 \times 2} = \frac{2}{6}$. So $\frac{1}{3} = \frac{2}{6}$.

12 $\frac{3}{6} = \frac{1}{2}$, which is NOT 1. The others all have numerator = denominator.

13 $\frac{1}{2} = \frac{3}{6}$. Since $\frac{2}{6} < \frac{3}{6}$, it is less than $\frac{1}{2}$.

14 $\frac{3}{6} + \frac{2}{6} = \frac{5}{6}$ cup.

Find more at
ViewMath.com/VA-Grade3

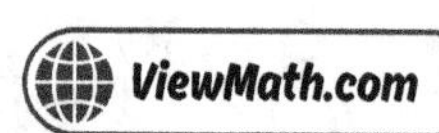

15 $\dfrac{4-2}{6} = \dfrac{2}{6}$.

16 $3 \times 3 + 1 = 9 + 1 = 10$. So $3\dfrac{1}{3} = \dfrac{10}{3}$.

17 When both hands point to 12, the time is 12:00 *(noon or midnight)*.

18 Each inch has 2 half inches. So $3 \times 2 = 6$ half inches.

19 $500 + 500 = 1{,}000$ g *(which is the same as 1 kg)*.

20 Total spent: $\$2.35 + \$1.80 = \$4.15$. Money left: $\$6.00 - \$4.15 = \$1.85$. Since $\$1.85 < \$2.00$, she does not have enough.

21 Higher numbers mean hotter temperatures. $72 > 45$, so $72°F$ is warmer.

22 Red: $8 \times 2 = 16$. Green: $5 \times 2 = 10$. Total: $16 + 10 = 26$ crayons.

23 A line plot uses X marks stacked above a number line to show data.

24 A coin has two sides — heads and tails. Each side has the same chance, so getting heads is equally likely as getting tails.

25 The top point where all the triangular faces of a pyramid meet is called the **apex**.

26 A unit square is a square with sides that are 1 unit long. We use unit squares to measure area.

27 Area $=$ length $\times$ width, so $42 = 7 \times$ width. Width $= 42 \div 7 = 6$ cm.

Find more at
ViewMath.com/VA-Grade3

28 The corner of a book forms a right angle, which is exactly 90°.

29 A nonagon has 9 sides and 9 vertices. The number of sides always equals the number of vertices.

30 Same shape (circle) but different sizes (2 cm vs. 4 cm), so they are NOT congruent.

✅ Practice Test 2 — Answer Key

1	$9,000 + 200 + 5$	2	D	3	$22, 24, 26, 28, 30$	4	100	5	D	6	851	7	D
8	B	9	B	10	C	11	C	12	B	13	C	14	C	15	B	16	C	17	C
18	$1\frac{1}{2}$ inches	19	3,000 g	20	\$8.75	21	D	22	D	23	C	24	Likely	25	D
26	C	27	48 sq cm	28	90 degrees	29	B	30	Yes, flipping does not change shape or size

💡 Time to Learn! 💡

Go through the explanations below, **especially for the questions you missed**.

Understanding why each answer is correct makes you a stronger math thinker!

👍 **Tip:** Circle any questions you got wrong, then read their explanation carefully.

📖 Practice Test 2 — Detailed Explanations

1 $9,205 = 9,000 + 200 + 0 + 5$. The tens digit is 0, so there is no tens term.

Find more at
ViewMath.com/VA-Grade3

ViewMath.com

2. The hundreds digit is 7. Since $7 \geq 5$, round up. $6{,}782 \approx 7{,}000$.

3. The even numbers between 21 and 30 are $22, 24, 26, 28, 30$. They all end in an even digit.

4. $10{,}000 \div 100 = 100$. There are 100 hundreds in $10{,}000$.

5. The hundreds digit is 7. Since $7 \geq 5$, round up. $7{,}777 \approx 8{,}000$.

6. $583 + 268 = 851$. Ones: $3 + 8 = 11$, carry 1. Tens: $8 + 6 + 1 = 15$, carry 1. Hundreds: $5 + 2 + 1 = 8$.

7. Always start adding from the ones column (the rightmost column) and work your way to the left.

8. $3{,}412 \approx 3{,}000$ and $2{,}678 \approx 3{,}000$. So $3{,}000 + 3{,}000 = 6{,}000$.

9. Think: $9 \times ? = 72$. Since $9 \times 8 = 72$, the answer is 8.

10. $9 \times 7 = 63$. The missing number is 9.

11. The denominator went from 2 to 6 (multiplied by 3). Multiply the numerator too: $1 \times 3 = 3$. So $\frac{1}{2} = \frac{3}{6}$.

12. $\frac{3}{3} = 1$, so $\frac{9}{3} = 3$.

13. $\frac{3}{4}$ is only $\frac{1}{4}$ away from 1, which is the closest of all the choices.

14. $\frac{3}{8} + \frac{4}{8} = \frac{7}{8}$ mile.

15. To subtract fractions with like denominators, subtract the numerators and keep the denominator the same.

Find more at
ViewMath.com/VA-Grade3

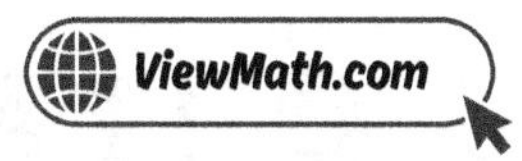

16 $2 \times 2 + 1 = 4 + 1 = 5$. So $2\frac{1}{2} = \frac{5}{2}$ cups.

17   "Half past" means 30 minutes after the hour. Half past 4 is 4:30.

18 $6 - 4\frac{1}{2} = 1\frac{1}{2}$ inches.

19 $3 \times 1,000 = 3,000$ g.

20   Cents: $40 + 35 = 75$. Dollars: $6 + 2 = 8$. Answer: $8.75.

21 $92°F$ is a hot summer temperature, perfect for the beach.

22   Lily forgot to multiply by the key. Each star = 2 kids, and there are 4 stars: $4 \times 2 = 8$ kids.

23   Line plots are best for measurement data. They show each individual measurement as an X on a number line, making it easy to see how values are spread out.

24   6 out of 8 sections are green. Since most sections are green, landing on green is likely.

25   A sphere is completely round with no flat faces, no edges, and no vertices.

26   5 rows $\times$ 6 columns $= 30$ unit squares. The area is 30 square units.

27   Area $= 6 \times 8 = 48$ sq cm.

28   A right angle is exactly $90°$.

29   Triangle (3), quadrilateral (4), pentagon (5), hexagon (6) — this is in order from fewest to most sides.

Find more at
ViewMath.com/VA-Grade3

30 Congruence only requires the same shape and same size. Flipping, turning, or sliding a shape does not change congruence.

Practice Test 3 — Answer Key

1 C	2 400	3 B	4 C	5 6,000	6 C	7 A	8 B
9 C	10 B	11 B	12 5	13 B	14 C	15 $\frac{2}{6}$	16 D
17 D	18 B	19 A	20 B	21 55°F	22 C	23 A	24 B

25 (a) A cereal box or brick. (b) A basketball or globe. (c) A can of soup or a drinking glass.   26 C

27 D   28 C   29 C   30 C

💡 Time to Learn! 💡

Go through the explanations below, **especially for the questions you missed**.

Understanding why each answer is correct makes you a stronger math thinker!

👍 **Tip:** Circle any questions you got wrong, then read their explanation carefully.

📖 Practice Test 3 — Detailed Explanations

1 Each place is 10 times bigger than the place to its right: ones → tens → hundreds → thousands.

2 The tens digit is 2. Since $2 < 5$, round down. $423 \approx 400$.

3 Odd + Odd = Even. For example, $3 + 5 = 8$ (even) and $7 + 9 = 16$ (even).

4 $9{,}999 + 1 = 10{,}000$. This is the first 5-digit number.

5 The hundreds digit is 5. Since $5 \geq 5$, round up. $5{,}500 \approx 6{,}000$.

6 In $367 + 485$, the ones column is $7 + 5 = 12$, which is 10 or more, so you must regroup. The other problems do not need regrouping.

7 No regrouping needed. Ones: $3 + 2 = 5$. Tens: $0 + 4 = 4$. Hundreds: $1 + 6 = 7$. Thousands: $2 + 5 = 7$. The sum is $7{,}745$.

8 $724 \approx 700$ and $389 \approx 400$. So $700 - 400 = 300$.

9 Skip count by 5s: $5, 10, 15, 20, 25, 30, 35$. That's 7 jumps. So $35 \div 5 = 7$.

10 $8 \times 6 = 48$. The missing number is 6.

11 $\frac{4}{6} = \frac{2}{3}$, $\frac{6}{9} = \frac{2}{3}$, and $\frac{8}{12} = \frac{2}{3}$. But $\frac{6}{8} = \frac{3}{4}$, which is not equivalent to $\frac{2}{3}$.

12 $\frac{2}{2} = 1$, so $\frac{10}{2} = 5$.

13 With the same numerator, the bigger denominator gives **smaller** pieces. $\frac{1}{3} > \frac{1}{8}$.

14 $\frac{1+5}{6} = \frac{6}{6} = 1$ whole.

15 $\frac{6}{6} - \frac{4}{6} = \frac{2}{6}$ of the bar.

Find more at
ViewMath.com/VA-Grade3

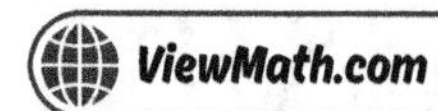
ViewMath.com

16. A proper fraction has a numerator smaller than the denominator. $\frac{2}{3}$ has $2 < 3$, so it is a proper fraction.

17. The short hand past 3 means the hour is 3. The long hand on 9 means $9 \times 5 = 45$ minutes. The time is 3:45.

18. Max started at 1 instead of 0, so the real length is $7 - 1 = 6$ inches.

19. Divide by 1,000: $9,000 \div 1,000 = 9$ kg.

20. $\$7.50 - \$4.25 = \$3.25$.

21. $80 - 25 = 55$. The nighttime temperature is $55°F$.

22. The key tells you what each picture or symbol stands for, such as "each star = 2 votes."

23. Each X mark represents one pencil. 3 X marks above $1\frac{1}{2}$ means 3 pencils measured $1\frac{1}{2}$ inches.

24. There are 6 green and 6 red — the same number of each. So picking green is equally likely as picking red.

25. Many everyday objects match 3D shapes: boxes are rectangular prisms, balls are spheres, and cans are cylinders.

26. $7 \times 3 = 21$. The garden is covered by 21 square tiles.

27. $6 \times 5 = 30$, $3 \times 8 = 24$, $4 \times 7 = 28$, $2 \times 10 = 20$. The 2×10 rectangle has the smallest area of 20 sq cm.

28. A line segment has two endpoints, so it does NOT go on forever. Lines and rays go on forever.

29. A pentagon has 5 sides. $5 + 3 = 8$ sides. An octagon has 8 sides.

Find more at
ViewMath.com/VA-Grade3

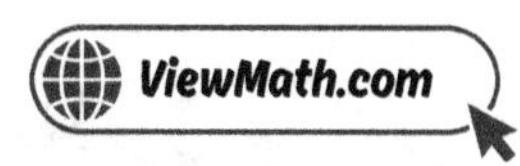

30. For shapes to be congruent, ALL side lengths must match. The third side is different (5 cm vs. 6 cm), so they are NOT congruent.

Great job checking your work!

Keep practicing and you'll be a math star!

Find more at
ViewMath.com/VA-Grade3

 ViewMath.com

9 798889 599151 0